LOUIS SHAW

Pioneer of  *e*

by P

FOREWARD

by Dr Mike Taylor (Editor 7¼ Gauge Society News)

The 7¼ Gauge Society Committee members including myself are pleased to commend and present this account of a true pioneer of our gauge. There is no doubt Louis Shaw must be remembered as a father of 7¼ Gauge commercial section, rather like Sir Arthur P Haywood is of the miniature gauge railways. We are indeed indebted to Peter James for his diligent and persistent research into the life and work of Louis Shaw. Peter James was amply supported by Simon Townsend and other contributors, and are all to be congratulated for making this historical record possible, amply illustrated with such wonderful photographs. The society is delighted to publish this work, not only for the benefit of our members but for all those interested in small railways. Further copies of this book will be for sale elsewhere and also from our secretary Bonnie Whisstock at 149 Surbiton Hill Park, Surbiton, Surrey, KT5 8EJ, at the published price.

A Publication of the 7¼" Gauge Society

Contents

Published by the 7¼in Gauge Society at 149, Subiton Hill Park, Surbiton, Surrey KT5 8EJ

ISBN 0 9514101 0 5 Louis Shaw – Pioneer of the 7¼ Inch Gauge.

Printed by Rainbow Advertising & Marketing Ltd, Cheltenham.

Chapter One Introduction

7¼ in gauge railways seem to be more popular today than ever before; but when Louis Shaw built his first 7¼ in locomotive back in 1910, such things were few and far between. Readers may wonder why we have gone to such lengths to record Louis' model engineering, and indeed I only came across his story by chance. Looking through some old railway books and magazines, I spotted a photograph of LORNA DOONE, credited as being built by 'Mr Lewis Shaw of Ilkeston'. With Ilkeston just a few miles away, I thought I would try to find out more. After all, a bit of local research could not take more than a few weeks; that was three years ago.

Many readers will probably never have heard of Louis Shaw before now. That is understandable, since there have been only the briefest of references to him in print, and many of those have mispelt his name! I was fortunate to make contact with Louis' family at an early stage, and to be shown the magnificent photographs which have been handed down, all of which are reproduced here. Between then and now I have received all sorts of generous assistance with my researches, from railway enthusiasts, historians and local people alike.

This booklet is intended to be a lasting tribute to Louis' model engineering career. I have tried to make it as comprehensive and as accurate as possible, but you readers must pardon any weaknesses, as most of the story all happened many, many years ago. Even after three years of research, I still do not know for sure just how many 7¼ in locos Louis built. In the end, it is better to share an incomplete record than to lose it all. For a year I was aided in my researches by Louis' son Godfrey, though due to his untimely death in 1986, he was unable to see the major breakthroughs, which occurred long afterwards. Although not a model engineer in his own right, he had worked on most of the engines with his father, and he was guardian to the last of R D LAWRIE, Louis' final locomotive. It would be unfair to underestimate his involvement with the locos, or his excellent memory for detail.

Before commencing a detailed story, it seems appropriate to set the scene by giving a brief account of the origins of the 7¼ in gauge in the U.K. Locomotive models to 1½ in scale have been built occasionally since the earliest days of railways, to demonstrate particular designs intended for the full size. In the early 1900s, 7¼ in was proposed by Henry Greenly to the Society of Model and Experimental Engineers, and was duly adopted by them as a standard, so that members using it would be able to run on other lines as well as their own.

Messrs Bassett-Lowke (B-L), at Northampton, were the first to build a 7¼in locomotive commercially. In 1908 they received an order from Mr E S Coats, of Paisley, to build a 7¼in gauge IMMINGHAM. Henry Greenly was commissioned to design the locomotive, which was completed amid as much publicity as B-L could muster. Soon, their catalogue featured a range of 7¼in locos, rolling stock, track and equipment. W J Bassett-Lowke quickly realised that something simpler, cheaper, was needed, so Greenly designed an 0-4-2, capable of traversing 25ft. radius curves. The first of those was sold in 1909 to Mr H F Gerhartz of Bradford, who built a simple line up and down his garden.

Meanwhile, Greenly had started building a third design, for a LNWR 'Precursor'. This was able to use the same cylinder block as the 0-4-2, and axleboxes were standard on all three locos. Once completed by Bassett-Lowke, this loco was sold to Mr Guy Mitchell of Sheffield. Mitchell's 'Brook House Model Railway' managed to fit an ingenious layout into a small space, part of it dual gauge with 3½in, and all fully signalled. By the time of their 1911 catalogue, B-L were also offering the 'Precursor' finished instead as GEORGE THE FIFTH, in order to keep in step with their full sized counterparts. Nevertheless, one speculates that sales were not that great, because B-L decided to mount a major display of their garden railway equipment at Olympia, for the Childrens Welfare Exhibition in January 1913. Over 8000 people were carried on the 7¼in gauge line in the 11 days of the event. GEORGE THE FIFTH went on to become B-L's most popular 7¼in gauge locomotive. Complete or as castings, it was still being offered in their catalogues as late as 1939.

Louis Shaw is one of only four British model engineers known to have used the 7¼in gauge prior to 1914. With his partner, he operated a railway in a public park from 1915. Not a demonstration garden railway, but a commercial miniature railway, this was the forerunner of the many such 7¼in gauge lines today. But oddly enough, Louis himself was a very quiet, shy man who always under-valued his own work while praising everyone else's. Later in his life he was interviewed by the Stanton Company magazine THE STANTONIAN, but was almost dismissive of his own achievements, though his interest in LORNA DOONE and tracing his lost engines never diminished.

Between 1910 and 1953, Louis designed and built at least seven 7¼in gauge locos, and what a career these themselves have led! BONNIE DUNDEE was the first locomotive on Kerr's Miniature Railway in Arbroath, a line which is still going strong today, some 53 years later. LORNA DOONE was the locomotive which first prompted the building of the Hilton Valley Railway, a line which itself lasted for 22 years, and is still fondly remembered around the Midlands. EUREKA was the first decent locomotive to arrive on the Greywood Central Railway; now at the Great Cockrow Railway it is still in regular use, a real tribute to Louis' sturdy workmanship back in the 1920s.

So we begin the story of Louis Shaw, a true pioneer of the 7¼in gauge.

Upper photo opposite:
7¼in gauge the Bassett-Lowke way; Guy Mitchell's railway at Totley, circa 1912.

Lower opposite:
Louis Shaw with his good friend Herbert Ballington and their families. Louis is standing in the centre, while Herbert is seated second from the left.

Louis Shaw with some of his early models on display, prior to 1910.

Steam Up! The 5in gauge line has been set up in the back yard, using GNR 277.

Chapter Two Longfield Lane - The First Lines, from 1910

Who was Louis Shaw? He was born on 1 November 1881 in Ilkeston, Derbyshire, an industrial mill town surrounded by collieries, and near to the Stanton-by-Dale ironworks in the Erewash valley. The Great Northern and Midland Railway lines were nearby, as the town lies midway between Derby and Nottingham. He began building models in 1902, at the age of twenty one. The first photograph that we have shows him with a miniature pit headstocks, a fairground wheel, and several model locomotives. By circa 1908 he had built GNR 277, a 4-2-2 tender loco which we believe was probably of 4¾in gauge. At that time, most small gauge tracks were laid at ground level, so it seems that Louis showed his initiative at that early date by building an elevated track and a special driving truck for it. Evidently though, 4¾in only whetted his appetite for greater things, and he turned to 7¼in gauge.

By 1910, Louis was living at 'The Mount', Longfield Lane, in Hallam Fields, midway between Ilkeston and Stanton-by-Dale. At the time the lane passed mainly through fields, though mushrooming housing development has rather taken over since then. 'The Mount' remained as Louis' family home until his death, and still stands to this day.

Louis' first 7¼in gauge loco was "G M R" No. 1910, based on a Midland Railway Compound, though itself a two cylinder 4-4-0. Although a good resemblance, it was clearly designed from the outset for work, having the driving wheels slightly less than scale, and the boiler slightly larger. A test track was laid for it through yard and garden at the back of the house. It pulled a simple but practical sit-astride carriage, and evidently the whole thing was a complete success.

By 1912, a longer line had been built along the length of the field next door to 'The Mount', 100 yards or more from end to end. This time, much heavier track materials were used; the chairs and rail were very probably purchased from Messrs Bassett-Lowke. As the field was for from flat, the lower sections of the line were built up on trestles to give an acceptable gradient. Halfway along the track was a signal, operated by a lever, or automatically by a ramp set between the rails. By this time, Louis had (we believe) repainted No. 1910 in authentic Midland Railway livery, as No. 1043. (The full size 1043 was a Midland Railway Deeley Compound built in March 1909.)

The original line of 1910 was built primarily to test the locomotive, although of course family and friends were invited to bring their children for a ride. Local people recall that the later line was a little more organised, with nominal charges made for rides. It proved to Louis that his little 4-4-0 could stand up to regular work, whilst the novelty of such rides could provide some additional income. Soon, the opportunity arose for him to move his railway onto a public site, literally just at one end of Longfield Lane! Fate had taken him by the hand.

Louis' first 7¼in gauge track was built in the field, outside the garden wall at Longfield Lane. His love of children is self-evident in this view, circa 1910. This faded photo is the first of all showing Louis' 7¼in locos in steam.

On Parade: Louis felt proud enough of his first 7¼in loco for it to be displayed.

1912, and bigger things! This was Louis' second line in the field at Longfield Lane, this time using Bassett-Lowke trackwork. Louis' good friend Herbert Ballington and his son are the passengers.

1912, further down the track, running close to the lane. Louis' home in the background. The semaphore can just be seen again in the distance.

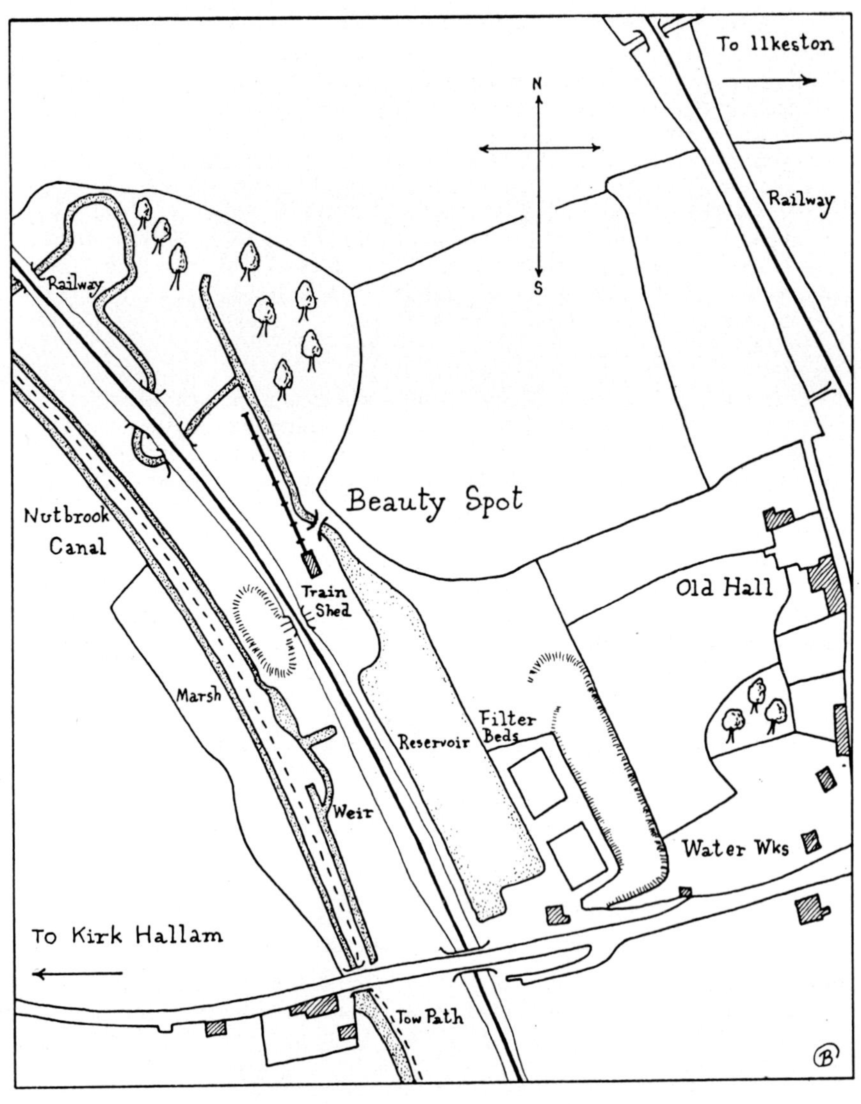

Based on Land Use Guide of 1914.

Chapter Three 'Beauty Spot' Days, 1915 to 1923

Little Hallam reservoir once supplied all the domestic water for the town of Ilkeston and its environs, but by about 1900 it was no longer needed. Instead, local anglers of enterprise proceeded to stock the man-made pools with fish, and after a few years it became the established and renowned site for anglers from miles around. On 10 February 1914, THE ILKESTON ADVERTISER carried a brief report that the whole pocket of land forming the reservoir and waterworks had been purchased by Mr S P Derbyshire of Nottingham, and he now donated the whole site back to the people of Ilkeston as a public amenity. Later in the year there were plans to install a zoo there but, after Mr Derbyshire had bought two polar bears for it, the Council changed their minds and opted for a more modest proposal.

The site itself was bordered by roads on the northerly and southerly boundaries, and by mineral railways to local collieries on the other two sides; these were branches off the Erewash Valley line, near Stanton-by-Dale Ironworks. It sloped steeply from north to south; the lower section was wooded although some formal gardens were laid out. Central to this was the larger pool, near to the old waterworks buildings, which became the boating lake. In both sections there were grassed areas for games and picnics; the upper section was mainly laid out in terraces with formal flowerbeds. Developed wholly from local funds, the site quickly became popular with folk from Ilkeston, Nottingham and around, who came to the gardens at weekends and holidays for fishing, boating, picnics, games and so on. They dubbed it the 'Beauty Spot'.

In 1915, Louis Shaw went into partnership with his close friend Herbert Ballington. Together, they dismantled the railway in the field at Longfield Lane, and before the Easter holiday they had relaid it at the 'Beauty Spot'. The new line was just over 100 yards in length, using the Bassett-Lowke track, but properly ballasted. It was a simple straight run parallel to a stream, from the boating lake jetty down towards the middle pool. At the jetty terminus, a wooden shed was built, with an extension into which the stock was shunted back each night for storage. George Barlow saw this when he rode on the line in 1923, and was able to describe it quite vividly.

In 1918, Louis completed a new, bigger locomotive to take over most of the work from No. 1043. This was a 4-4-2 of his own freelance design, based broadly on a Great Central Railway 'Jersey Lily'. Being much more powerful, the 4-4-2 was able to pull a heavier train, and Herbert Ballington built three very sturdy carriages, each with four seats slung in a well between the bogies.

During the 1920s, ownership of the land at the 'Beauty Spot' passed into the hands of the Ilkeston Co-operative Society. From then on the rental for the line was fixed yearly, based on the takings for the preceding August Bank Holiday weekend. The two partners viewed this with mixed feelings, since that weekend was always the best of the season, which could otherwise be poor if the weather went against them. They began to consider possible sites elsewhere. The August Bank Holiday of 1923 saw them very busy, but ironically this set a high rent for 1924, so they decided to make their move, and the line was lifted.

An early photograph at the 'Beauty Spot'.

This and the following two photographs were all taken over August Bank Holiday 1923, when the life of the 'Beauty Spot' line was coming to an end. This is a public train, and the coach design is shown clearly. Three of these very practical vehicles were built by Herbert Ballington circa 1920.

A classic portrait of Louis Shaw's third locomotive, built in 1918.

Here, Godfrey Shaw is at the controls as an admiring crowd stands behind.

1924, with a new locomotive on a new railway, on Mablethorpe sands.

A short train, with the family on the first carriage.

Chapter Four Sand and Steam, Mablethorpe, 1924-5

Louis Shaw and Herbert Ballington doubtless left the 'Beauty Spot' with some regret. It had been a convenient site close to home, though known only to local people. They had gained invaluable experience from running the railway there, but must have felt that it was time for greater things. In fact this marked the beginning of a new era for them, an era of new achievements. They worked hard during the winter of 1923-24, and before the Easter Bank Holiday a new railway had been constructed on the sands at Mablethorpe.

Back in the 1920s, Mablethorpe was a popular seaside resort, set on the Lincolnshire coast half way between Skegness and Cleethorpes. It was served by a loop line of the Great Northern Railway/LNER Lincolnshire coast main line, Peterborough to Grimsby. From the station it was a short walk along High Street to the sands. Close by the shore the Central Promenade stretched to the south whereas a number of caravan sites and beach chalets were situated along Quebec Road, to the north.

At the time, two family businesses were influential in the town, the Parrys and the Parrotts. The Parrott family had the donkey concessions on the sands, and several restaurants and cafes along Quebec Road, north of the main amusements which were beside the Central Promenade. Mr Parrott was well known to several Ilkeston people, and it seems that he persuaded Shaw and Ballington be build their 7¼in 'model' railway on the sands at this northern end of the resort. The site itself was next to the dunes near where Golf Road joined Quebec Road, and very close to one of Mr Parrott's cafes.

The same simple track was used, but to add authenticity a few scale signals were built, and halfway along the line a mock tunnel was constructed using a wooden frame covered with canvas. This also came in useful as a lockup shed in which to store the locomotive and stock at night. On occasion, when a member of the two mens' families came to visit, the tunnel was used as sleeping quarters as well, or else a chalet would be rented. Mr and Mrs Pearson at No 1 Golf Road became good friends of the Shaws and would help out with supplies like drinking water.

Godfrey Shaw recalled that the 4-4-0 No 1043, and the 1918 4-4-2 were used here on occasion, but the principal locomotive was a new 4-4-2 which Louis completed in time for the line's opening, it was of similar design to the first one. Louis always drove the loco, whereas Herbert would collect the fares and look after anything else that needed doing. Later, in 1925, Louis completed his first 4-6-2, PACIFIC, which also made a brief appearance here, remaining until the line was lifted.

The new line brought mixed success. On the one hand, the two partners had clearly been wise to move their railway to an established resort. However, they had badly underestimated the problems entailed having laid the track directly onto the sands on an unprotected beach. Many friends, including Mr Anthony, remember that there were often mornings when it was found that the line was completely buried by sand. Herbert would have to dig it out again before they could begin running; occasionally they needed to do this two or three times in a day! As if that was not enough, the sand got absolutely everywhere. The bearings on the locomotive and stock quickly wore down, and it took all Louis' efforts in his Ilkeston workshop over the closed season to keep things in running order.

At the end of the 1925 season, approaches were made to Louis and Herbert to move their railway onto a new site in the centre of Mablethorpe. It was next to the High Street, where virtually every visitor passed on their way from the main line station to the sea front. They put down a temporary track there for a short time and it was a considerable improvement; a more popular railway and much easier to operate. This experiment opened the way for the creation of the finest railway of the Shaw-Ballington partnership.

This is PACIFIC with its train, in use on the temporary track next to Mablethorpe High Street late in 1925.

1924, and the second Atlantic is posed for the record. Compare with the earlier locomotive on page 13.

The end of the day, with the tunnel and also a semaphore in view.

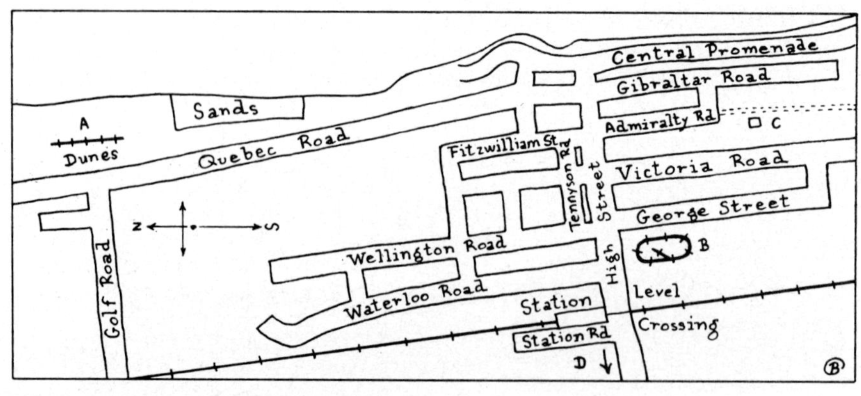

Key

A: The Railway on the sands, 1924-5
B: The Mablethorpe Model Rly, 1926-39, see detailed map
C: Spire House, H.K.'s home
D: Willoughby House, H.K.'s earlier home

Sketch Map of Mablethorpe

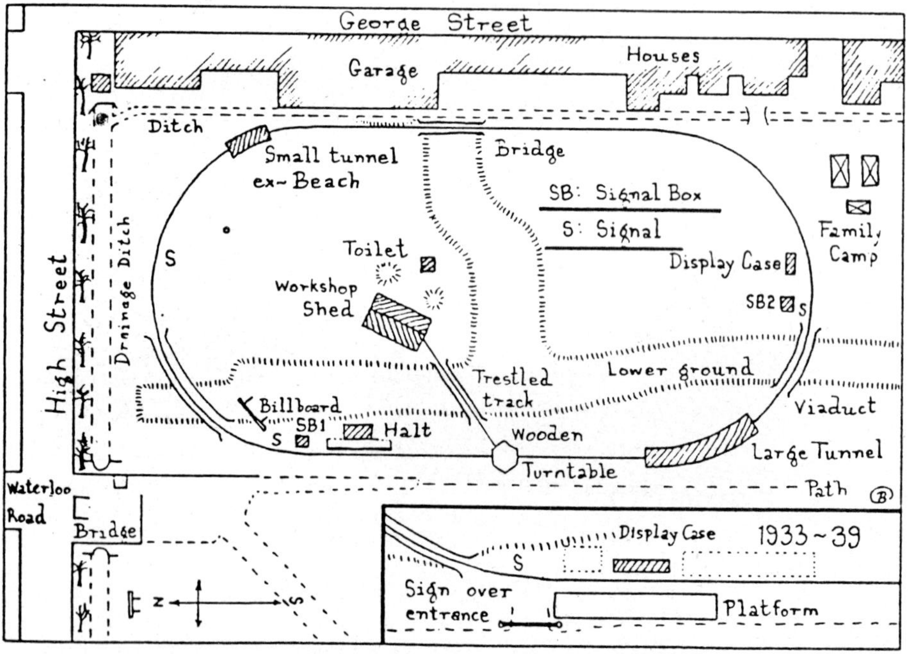

The Mablethorpe Model Railway

Chapter Five The Mablethorpe Model Railway, 1926-1939

The centre of Mablethorpe was still being developed in the 1920s. The Parry family owned a brickworks on the outskirts of the town and most of the land between there and George Street, though they rarely came near the town themselves. Instead, their business affairs were managed by their agent Mr Seed, or later Mr Negus. Through the 1920s and 1930s there was development for houses and shops as the resort became more established. Between George Street and High Street was a large field. For some time, this land had been divided into sub-plots and leased to showmen and other traders for amusements, fortune tellers, kiosks and so on. They kept their caravans in another section of the field for living quarters during the season.

Through their railway on the sands, Louis Shaw and Herbert Ballington had come to know one Percy Harding-Kiff of Willoughby House, Mablethorpe. 'H-K' had a shrewd eye to see where there was money to be earned, and must have looked upon the model railway as potentially a very good business, especially at the town centre site. Such was its success there that in 1926 H-K took out a 20 year lease on the land, and entered into partnership with Louis and Herbert.

Now with a prime site all to themselves, the two partners took their opportunity to instal a 'proper' railway, in so far as space allowed. The entrance to the field was by a concrete bridge over a drainage dyke, so they built their model railway station next to this, on the opposite side from the other amusements and stall holders. Throughout the life of the line there was a series of large wooden framed billboards aimed to attract would-be customers for a ride on the train.

The station itself was marked by a wooden platform and a shelter, both on the inside of the track, which was on oval about 300 yards long. Bullhead rail was used, with keyed chairs; and over one weekend semaphore signals were installed operated by electric motors which were controlled automatically from a small signal box next to the station. Around the circuit were a number of dips in the ground. Two of these were bridged by viaducts with trellis fences, whereas on the straight along by the back gardens of George Street they built an impressive wooden girder bridge. There were also two tunnels, the smaller of which was relocated from the railway on the sands. In the centre of the circuit was the wooden engine shed, with a small workshop in it where later Louis worked on gauge 0 and 1 models for H-K. The shed was reached by a turntable from the main line, and then across some trestles.

The main locomotive here from the beginning was Louis' first freelance 4-6-2, PACIFIC. This name was carried on the loco's centre splasher, although it was also known by its nickname (never carried) of LORNA DOONE. Herbert had built a second set of carriages. By 1932 there were 10 carriages in all, six of them were 6ft long carrying two passengers each, the other four 6ft 6in long carrying two adults and two children each. Most were made up into articulated pairs, two pairs being a normal train. At most times one locomotive in steam was sufficient, although on peak days two trains were used, one loading in the station while the other was going round. Initially one of the Atlantics was used as second engine.

Louis Shaw and Herbert Ballington, with PACIFIC and a full train. The houses in the background are in George Street.

This view shows all the details of the proper station which was built once the line became permanent.

Two views taken on August Bank Holiday 1926. Here, Louis is at the controls of an Atlantic. High Street is in the background.

Here, Mr Stones senior is holding his younger son, also his brother. Herbert is punching tickets in the background. Note the seating arrangement.

In the early days, Louis always drove the locomotive and Herbert took the money, just as they had worked before. For 2d they also sold souvenir photographs of the train; 'send your friends one' proclaimed a sign on the station. By 1929, PACIFIC had been joined on the regular service by a very similar locomotive, but painted in authentic LNER livery carrying the number 4471. It was about this time when Harding-Kiff started to take a closer interest in the model railway. During the season, Louis would stay at Willoughby House, H-K's home, where a workshop for larger gauge models was set up in an out-building. Until then most of the engineering had still been carried out at home in Ilkeston. H-K had bought a Bassett-Lowke GEORGE THE FIFTH from somewhere and kept it at the railway. However, it would only pull a third of the load that the 4-6-2 could manage, and no photographs have come to light showing it in service. In July 1931 he advertised it for sale in the MODEL ENGINEER (M E).

H-K came to realise that Louis' model engineering skills were an asset in themselves. Through the pages of the M E he began to buy and sell locomotive models up to 7¼in gauge, and found that here too there was money to be made. Louis was to spend much more time in the workshops instead of being able to drive the locos. He must have viewed the change of emphasis with regret, but he had little choice in the matter. His good friend of old Herbert Ballington had retired from the partnership, and rather than give up all that he had worked for, Louis accepted H-K instead, though he soon found himself rather more of an employee than a partner.

One of the first effects of the change was that Louis' early locos now had to be restored to use, or sold. In May 1932 the Mablethorpe Miniature Railway was described in the M E, and a new locomotive had recently been added to stock - a G. C. R. Atlantic, presumably in fact one of the earlier two freshly overhauled. There were improvements at the railway as well. Over the winter of 1932-33 the old station was demolished and a new concrete platform put down on the outside of the track nearer to the viaduct. A big new sign was made up which announced MODEL RAILWAY, THE ONLY NOVELTY IN MABLETHROPE. It was probably true! Opposite the platform they put a display case which often featured small locos that H-K had for sale. One annual visitor to the line, Geof Nicholson, remembers a big LBSCR tank loco ABERGAVENNY along the back of the case, and others including a blue GER 0-6-0. H-K also managed to squeeze a small sales office into one end of the loco shed.

Louis' other Atlantic was also over hauled, and emerged with a Belpaire firebox carrying a new livery as LNER No 1933. During 1935 no less than five different 7¼in gauge locos appeared on the line at different times, four of them built by Louis, and three of them advertised for sale by H-K in the M E of 1 August that year. In order to save unnecessary confusion (!) these will be described in Chapter 7. Evidently H-K was very pleased to have Louis' practical 4-6-2 design working the public service. He calculated that the loco had run 10,000 miles, and wrote in to the M E to that effect. A supplementary reference appeared in the M E of 30 January 1936 which is worth quoting in full for its interest, although parts of it do seem very hard to believe. H-K was a skilled showman and never lost any opportunity to publicise the railway.

Above: A classic view in the station.

Photograph overleaf: Early days at the 'Beauty Spot'; with family and friends in this pose. Godfrey Shaw is sitting with his father.

Circa 1932, with LNER 4471 now in regular use. The coaches have been articulated into double sets. Louis is too engaged in the workshop to drive himself; we believe that this driver may be Percy Harding-Kiff.

This view is no later than 1932, because the signal box is still present.

A Good Loco. Record.

> 'Since my note on the passenger carrying miniature railway at Mablethorpe appeared, Mr P. Harding-Kiff sends me the following particulars:- "With reference to the recent note in 'Smoke Rings' concerning our 1½in scale 7¼in gauge 'Pacific' this record has been spread over a period of the past ten years, I should like to add that it was built by a great amateur, Mr Louis Shaw, of Ilkeston, and it still has its original wheels, cylinders and motion. Certain wear of bearings has had to be taken up and new pins fitted, but apart from this, it is to-day as good as ever and still hauls three tons with ease. Judging by its present condition it will be working in another ten years' time. It frequently travel s 60 actual miles in one day's running, hauling 3 tons most of the time."'

Upon consideration, it would appear that the claim in the last sentence was impossible! All the stock on the Mablethorpe Model Railway coupled together would not have made a three ton load; and allowing for loading times it would have taken about 24 hours to cover sixty miles on the circuit!

Only a matter of months later, in mid 1936, Louis completed his last 4-6-2. This was to the same freelance design as the earlier ones, but carried LNER livery as No 4472. On 20 June 1936 H-K moved home, into Spire House; it took some time to move the workshops into the outbuildings there. But Louis had had enough of being ordered about, and sold out his share of the partnership. After all his model engineering right back to 1910 and before, it must have been a hurtful insult for H-K to have described him as an amateur in the 'Model Engineer'. Louis lost heavily in the deal, and that was a source of ill feeling towards H-K on the part of all Louis' friends in Ilkeston for some time afterwards. Nevertheless he took what he was offered, and retired away from his model railway back home to Ilkeston.

The railway was given a facelift for the 1937 season with newly painted structures and flowerbeds. In June 1937 several paragraphs on the line were published in THE RAILWAY MAGAZINE following a visit by Charles H Lea. Mr Lea reported four locos present: two freelance Pacifics in LNER livery, a GEORGE THE FIFTH (another one!), and a G.C. type Atlantic. The line continued to run as before, until the outbreak of war in September 1939 when it closed, and the stock was taken away to Spire House for storage.

The track remained in situ until 1940 when the field was requisitioned for use as a tank disposal point. H-K tried to object to this, but the Major in command replied by indicating he might need to drive his tanks all over the track! In the event a compromise was reached, and the trestle by the shed was left intact, while the army lifted the circle of track and stacked the panels neatly in one corner of the field.

In 1947, after the war was over, H-K held a big sale of most of the models which he had kept in storage. The lease for the old site of the railway came to an end, and it was redeveloped with what is now Seacroft Road being built right across it. H-K continued to deal in locomotives periodically through to the mid 1960s when he returned to his native country of Australia. Mablethorpe has seen at least three different miniature railways since the 1940s (and still has one today), but none of them in the same class as the Mablethorpe Model Railway built in 1926 by Louis Shaw and Herbert Ballington.

An unusual view taken in 1933. 4471 is just setting off for another circuit. In the background the tunnel can be seen, which was moved there from the earlier line on the sands.

The new station in 1933, with 4471 awaiting duty. Note the billboard sign, the workshop shed, and signs of domesticity – caravan and tents on the far right, and washing on the line to the left of the shed. The fares are also displayed.

Chapter Six Post War Days

After the end of his time at Mablethorpe, Louis Shaw was totally disillusioned and took a well earned rest from his model engineering endeavours. He returned home to Longfield Lane, Ilkeston, and secured a job working in the foundries at Stanton-by-Dale ironworks. However, this did not mean that he lost touch with his model building friends; quite the contrary. One good friend was Albert Barton, also of Ilkeston. Back in 1928, Albert had seen one of Louis' 7¼in 4-6-2s, and admired it so much that he decided to build one himself. Louis kept in close touch with Albert's efforts, helping to build the loco's boiler, and lending him patterns when they were needed. By the mid 1930s Albert's loco. was complete, and he used it locally on a portable track for fetes and special events, such as the Shipley Gala each August bank holiday. After that loco. was completed, Albert and his son Dennis started slowly working on a new freelance 4-6-2 which was to become No 5065 ALBERTA, now PRINCESS ELIZABETH on the Walsall Steam Railway.

In 1940 Louis decided to build one last 7¼in gauge locomotive. In essence he planned it from the beginning to be his swansong, and it took him thirteen years to complete. It was a 4-4-2 close to an accurate reproduction of a GNR large Atlantic.

During the years 1945 to 1950 a test track was installed in Tathams Mill, Ilkeston, for the 'Needlemakers Arms' group of model engineers, over which Stanley Battison was a major influence, and Louis was a regular visitor there. (Battison himself built quite a number of large miniature locomotives; see the article on him published in the Heywood Society Journal, Autumn 1987 issue.) However, on 23 January 1948 THE STANTONIAN announced the formation within the Stanton Ironworks of a society of model engineers and craftsmen. This became 'The Stanton Experimental Engineers', of which Louis was a founding member and very much the elder statesman. With his long experience, he was always the first to help anyone who might need his advice, and his influence over the new generations of model engineers in the area was immense.

In 1953 his last 4-4-2 was completed, and he named it R D LAWRIE in tribute to a close friend who had helped him out in the troubled times after his return from Mablethorpe. He built a short test track for it in his back garden, with a corrugated iron shed in which the loco. was normally stored. On occasion he also steamed it on site at the ironworks, where another test track had been built. By this time, Louis' son Godfrey had become disillusioned with model engineering after many days spent helping his father in the workshop and seeing the outcome of it all in the collapse of Louis' endeavours at Mablethorpe. However, instead Louis found that his grandson Brian shared his enthusiasm for such things. They would often travel together to exhibitions or special events.

By 1958 Louis had also become a good friend of Sam Monk, a factory owner who lived at Caythorpe, north east of Nottingham. Several years before, Sam had built a 7¼in gauge DUCHESS OF BUCCLEUCH, and in 1954 he had purchased the 10¼in ROYAL SCOT built by Stanley Battison. Now at his new home he built a circular track with dual gauge so that he could run these and his other locomotives.

At Caythorpe, circa 1960, with Louis Shaw's final 7¼in locomotive R D LAWRIE. Sam Monk has finally got his wish to drive. The passengers are Louis, his wife, daughter in law, and grandson, Brian.

Here, Sam's son watches the gauges, as Sam talks things over with Louis.

Louis helped with the construction of the track, along with Sam's other friends such as Stanley Battison, Wilf Machin and Cyril Boothby. In 1959 louis moved R D LAWRIE to Caythorpe, and steamed it there regularly. Louis must have welcomed the continuous track to run on, instead of having to shuffle up and down the back garden all the time.

Sadly, Sam Monk died suddenly in August 1963. Louis had to negotiate for some months in order to retrieve R D LAWRIE, because he had to prove his ownership of the loco. to Sam's executors. Once it was back it was rarely steamed. By this time, many of Louis' other friends such as Stanley Battison and Albert Barton had also passed away, so Louis lived a quiet life at home, save for the odd visit to shows. Except for two bright moments (recounted later), Louis had lost contact with his old 7¼in gauge locomotives. He died on 5 August 1971, at the age of 89. Godfrey Shaw died exactly fifteen years later on 5 August 1986, whilst an overhaul of R D LAWRIE was under way. Nevertheless Louis' locomotives live on, as we shall see.

Louis with one of his many smaller gauge models.

Chapter Seven The Story of the Locomotives

The following account is one which has been pieced together from all the memories and photographs which have been gathered during my researches. However, it must be said that this has been the most difficult of tasks, for whereas a railway being built may create a documentary reference, a lease agreement or a Council minute for example, such evidence is rare in the case of a locomotive. Nevertheless I believe that I have come to understand much of the story, and certainly there is nobody else left whom I have not questioned about it.

One thing that is clear is that Louis Shaw was a great traditionalist in the way that he built his locomotives. One of his principal machine tools right to the end was a Greyson Treadle lathe, and Godfrey Shaw nursed memories of powering the treadle many times in the 1920s and 30s as Louis would craft another component. Another factor beyond doubt is that Louis' engines were worked very hard, and their design and construction had this in mind. On several occasions, Louis started components for two locos together, completing the second one before the first was retired for overhaul. Hence as Godfrey recalled to me, there was always more than one loco. under construction at Ilkeston at any one time.

When he was interviewed in the STANTON AND STAVELEY NEWS in 1967, Louis was sure that he had built at least twelve locomotives in all. From this we can assume that he was referring primarily to 7¼in locos, though including some of the larger of his other models, such as the 4¾in 4-2-2. He must have excluded all his gauge 0 and 1 locos of which he built many at Mablethorpe for H-K.

I have still found quite some fascination with the story of Louis' 7¼in locomotives, incomplete though it may be. Perhaps the publication of this booklet will bring still more details to light.

The 4-4-0

Louis' first 7¼in locomotive was the Midland Railway 4-4-0 described in Chapters Two and Three. Although not a scale model this had many of the features of its prototype, including connecting rods being inside the coupling rod. Initially it appeared as G M R No 1910, then later as Midland Railway No 1043 Godfrey Shaw thought that Louis would never have restyled any of his locomotives, nor even repainted them in different liveries. However, despite careful scrutiny of Nos 1910 and 1043, I have not been able to find any substantial difference between them, and this has led me to believe that in fact they are the same loco.

This locomotive worked at Longfield Lane, then at the 'Beauty Spot', and possibly briefly on the sands at Mablethorpe. I know nothing if its later history, although a locomotive of similar description was purchased by Colin Cartwright in the 1960s from Bill Southern of Hinckley, Leicestershire. Colin didn't keep the 4-4-0 for long though, before exchanging it back for a bigger loco. Bill Southern carried out engineering work for Percy Harding-Kiff in H-K's later years as a dealer.

The Early 4-4-2s

Just like Henry Greenly in the larger 15in gauge, Louis quickly found that scale models gave rise to problems with steaming, power and maintenance. Louis' answer was his own freelance 4-4-2 design, inspired somewhat by a Great Central Railway style, but with a round topped firebox instead of a square topped Belpaire. leading dimensions were cylinders 2¼in by 3in, 8in driving wheels, 4½in bogie wheels and an overall length of 8 ft. Two injectors were fitted, along with steam and hand brakes, and sanding gear.

The first of these locos was brought into service at the 'Beauty Spot'. It carried the letters B & S Ry on the tender (Ballington and Shaw), and also clearly visible on the photos is a worksplate which read 'L. Shaw Ilkeston 1918', clearly dating the year when it was completed. This loco may have appeared briefly on Mablethorpe sands, after which it was in store for a while; unnecessary to overhaul it while the other 4-4-2 was in service. However, it was rebuilt circa 1934 with a square topped firebox and then it joined the working fleet on the Mablethorpe Model Railway. George Barlow saw this loco in the shed there on 9 June 1935. We know it was this loco he saw because shortly after the war, George met Percy Harding-Kiff again, when H-K was down at New Romney collecting a locomotive. They found that they had met before, when H-K had opened the shed for George at Mablethorpe back in 1935. George commented how the 4-4-2 he saw then was similar to the one he had ridden behind even earlier, in 1923 at the 'Beauty Spot'. H-K said straight away that indeed it had been the same loco on both dates.

In 1936 Ivor Gotheridge spent a holiday with his family in Hornsea, on the Yorkshire coast south of Bridlington. Set back from the promenade up a little slope was a small miniature railway which went around a boating pool. Ivor recalled clearly that the locomotive was a 4-4-2 of about 7¼in gauge, Great Central style with a square top firebox. This may have been Louis' 1918 loco. Quite possibly H-K could have hired it out for a season, because Mr Lea reported one Atlantic back at Mablethorpe in the 'Railway Magazine' of June 1937.

What became of the loco after that we do not know, except for a memorable event in the 1960s. Louis had come down with Brian Shaw to the annual Model Engineering Exhibition held in London. On a stand belonging to the Kodak Society of Experimental Engineers, they were astonished to find one of Louis' old 7¼in 4-4-2s, still apparently with its round topped firebox. The Steward did not believe at first that Louis had built it, but Louis was able to recall so many features of its construction that he proved beyond doubt that the loco was one of his. This episode greatly amused Brian Shaw, Mr Anthony, and other Ilkeston friends who were with him.

The Kodak Society of Experimental Engineers and Craftsmen was formed within employees of that company, and was based at Wealdstone near Harrow in Middlesex. However, the group was wound up in the late 1960s, and I have not been able to trace the 4-4-2 any further. Nor can I explain how the shape of the firebox had changed again, except that possibly the square top added to it at Mablethorpe was merely a dummy. If that was not the case then the above account is an amalgam of two different locos, which I regret is also a perfectly possible explanation.

Turning now to Louis' second 4-4-2, this made its first appearance on mablethorpe Sands in 1924. It had similar leading dimensions to the 1918 loco, but carried a different livery, with 'M Shaw & Ballington R' on the tender. This loco also acted as second engine during the early years of the Mablethorpe Model Railway, being described as recently added to stock there in the MODEL ENGINEER account of May 1932. In 1933 it was rebuilt with a square topped firebox, and acquired a new apple green livery as LNER No 1933. After that it continued as spare engine, until February 1935 when H-K advertised it for sale in the M E. It was purchased by one Matthew Kerr, of Balfield Farm near Dundee, where it arrived on 5 April.

Although then a dairy farmer by trade, Matthew "was a young man of very decisive railway enthusiasm". He put down a short track at his father's farm, but this was not enough for him. On the coast seventeen miles away is the town of Arbroath, and soon he had negotiated with the Council there to run a miniature railway in West Links Park during the summer months. He named the 4-4-2 BONNIE DUNDEE, although it was always known by its nickname of 'Tibby'. By whatever name, it proved to be an instant success at its new home. Throughout the season it seemed that the little train of six coaches was always full.

Kerr's Miniature Railway was lifted for the winter 1935-36, and 'Tibby' repainted at the farm as LNER No 1936, with a small plate 'Rebuilt KMR 1936'. For both seasons of 1936 and 1937 Matthew laid down his railway at Arbroath again, and each year the layout became more ambitious. In 1937 there were two stations both with turntables and signal boxes, 700 ft of track, and other authentic features like a footbridge.

The only problem with all this was that 'Tibby' was getting a bit small for the increased traffic. Matthew decided that he really needed to build his railway in a more permanent form using a larger gauge. He got in touch with John Thurston of Farnborough, who at that time had a 10¼in loco for sale, a 4-4-0 which had been built some years before by HCS Bullock. They arranged a part exchange deal; Kerrs Miniature Railway was relaid in 10¼in gauge, and in that form it is still running today.

Louis Shaw's little 4-4-2, now as LNER No 1936, did not stay long in Farnborough before it was sold again. John Thurston's daughter has recalled that it went to a railway at Kingston upon Thames, but we have not been able to find out any details about this. Quite some time later, in the 1950s and 60s, Mr N H Mann operated a 7¼in gauge garden railway at his home in Kingston, and he may have owned the 4-4-2 for a while. We do know that at least by the early 1970s this loco. was with Mr W A Locker, who owned a private railway at his home in Cuddington, near Tarporley in Cheshire. Mr Locker had a most interesting collection of locos, including the 7¼in Midland Railway 0-4-4 tank which is now on the Spinney Railway. He told some visiting enthusiasts that he had purchased this loco. at Harding Kiff's big sale in Mablethorpe right back in 1947, along with a GEORGE THE FIFTH. Quite possibly he could have purchased the 4-4-2 through H-K as well.

Mr Locker died in 1975, and his model collection was sold at auction by Messrs Lewis Taylor & Son of Stoke on Trent on 27 January 1976. The 4-4-2 LNER No 1936 was purchased by Mark Bamford, but he sold it some time later, and for the

This photograph was sent by Harding-Kiff to Matthew Kerr, advertising the loco for sale. Atlantic LNER 1933 is on one of the raised sections of the Mablethorpe track. Comparisons between this and the other Atlantic views are intriguing.

'Tibby' in use at Kerr's Miniature Railway, Arbroath, in 1936.

In John Thurston's yard at Propsect Road, Farnborough, in the early months of 1938. On the left is 'Tibby', 7¼″g, which has arrived from Arbroath in exchange for 'Gladstone' (10¼″g, centre), which was originally built by HCS Bullock.

PACIFIC on duty in 1926.

moment I have lost track of this locomotive. I am told that it was sold to a collector in the Preston area in 1983; perhaps some member of the 7¼in Gauge Society will know more?

The 4-6-2s

These locomotives are an intriguing bunch. I have spent many many hours talking over the details with my friends, and often we have ended up more confused than when we started! In the account that follows I have tried hard to distinguish facts which I know to be accurate from theories which I believe to be correct.

Along with other members of the family, Godfrey Shaw remembered that the first 4-6-2 was started before the completion of the 1918 Atlantic, and by the early 1920s there were two 4-6-2s under construction in Louis' Ilkeston workshop. We are not certain how his design came about, but Louis must have been aware of what locomotives had been designed for other gauges. Henry Greenly had published the outline design of the 15in gauge Class 60 4-6-2 by 1913. When one was ordered, he put Ernest Twining in charge of preparing detailed drawings, and in 1915 Twining himself published an 'improved' 4-6-2 design, this time for a 3½in gauge loco featuring Gooch valve gear. On the other hand I am inclined to think that Louis largely relied upon his own initiative to develop the 7¼in 4-6-2s. A Pacific offered a larger boiler and more weight, and there is no doubt from Louis' other achievements that he could have thought this out for himself, encouraged by seeing such plans in the MODEL ENGINEER.

What resulted were probably the most practical 7¼in locomotives of their day. Principal dimensions were weight 8 cwt, length with tender 9 ft, cylinders 2¼in bore by 3½in stroke, 9 in diameter driving wheels; Stephenson's link valve gear. Working pressure was between 60 and 100 lbs; at 100 lbs they would pull a load of 3 tons. High and low pressure injectors were fitted, along with steam brakes and sanding gear.

Louis' first 4-6-2 was PACIFIC, a distinctive locomotive in the photographs due to its freelance black livery. This featured MODEL RAILWAY on the tender, six boiler bands, PACIFIC on the centre splasher, and painted on the cabside were the words 'L Shaw Ilkeston 1925'. Unlike Louis' later 4-6-2s, PACIFIC had a rectangular hole in the top of its cab, through which its driver could grip the regulator. This locomotive ran briefly at the railway on the sands in 1925 (several people remember seeing the black engine there), and thanks to Ken Blackham, we have a photo of it running later in 1925 on the temporary line which proved the viability of the High Street site. Once the Mablethorpe Model Railway was established there in 1926, PACIFIC took the lion's share of the work in the early years. It was more commonly referred to by its nickname of LORNA DOONE.

What became of PACIFIC I do not know, but by circa 1932 it had disappeared from view. I have been told that it lay scrapped or dismantled at Mablethorpe. Louis provided the frames and boiler for Albert Barton's 4-6-2 which was being built at this time. Albert's family remember these being collected from Stanton Ironworks, where Louis worked. I believe that these may have been from PACIFIC. When he wrote in to the M E in January 1936 H-K claimed that the locomotive then had been working for ten years, which if true could have meant a rebuild of PACIFIC, but H-K was a good salesman, and would have been prone to a bit of exaggeration. In any case, local people in Mablethorpe have told me that they saw the black 4-6-2 side by side with the green 4-6-2 (the next one to appear).

I believe that the second of the two 4-6-2s which Louis built in Ilkeston was completed well before 1930, even though I have no record of it before 1935. It appeared at Mablethorpe in that year painted as LMS No 6200. Why Louis styled it as LMS I do not know, but the livery cannot have been applied before 1933, as the prototype locomotive was constructed by the LMS at Crewe in that year. Although of the same general design as PACIFIC, there are features of LMS No 6200 which most definitely distinguish it from all the others. It has a very simple six wheel tender. The boiler cladding does not cover the bottom of the firebox, and there it can be seen that the boiler is rivetted; the bottom of the firebox is not straight. There are only four rivets holding the smokebox saddle to the smokebox, which is shorter with the chimney mounted at the back. The shape of the frames up to the smokebox saddle are quite different from any other. I am quite certain that LMS No 6200 was not a rebuild of any of the others.

A number of Ilkeston people including Albert Barton recalled seeing one of Louis' 4-6-2s in use on portable tracks locally. I believe that those memories are of this loco. I do know that it did no work at all at Mablethorpe; H-K put it up for sale (M E 1 August & 24 October 1935). It was sold for £150 to Mr Harry J B Ferguson of Carnoustie, seaside town on the Angus coast westward of Arbroath. Carnoustie Town Council had seen the success of Kerr's Miniature Railway and thought that a similar miniature railway would make a good attraction. Harry Ferguson had recently retired, and knew a bit about such things, having built 2½in gauge locos. He had also helped out a bit on the Arbroath line. A councillor lived near Harry, and he was talked into building a railway along the Promenade, the 'Beach Model Railway'.

Matthew Kerr put Harry in touch with H-K, and LMS No 6200 was delivered by rail late in 1935. Later, when writing an article in the M E, Harry said that it "took a great deal of work to get the loco up to scratch". Once the work was done he named it CORONATION. He also had a set of coaches built, so it was 1937 before he managed to open his railway, which was a straight up and down run, rather boxed in between a childrens' play area and a paddling pool. The line worked until the outbreak of war, and Harry had nothing but praise for the loco; so much so that he began building a new 4-6-2 in his workshop based on Louis' design. On one day following an accident, No 6200 even worked as an 0-6-2, as Harry had taken a party booking so the show had to go on.

By the end of the war Harry Ferguson was in ill health, so he put No 6200 and the part built sister engine up for sale. No 6200 was purchased by Mr J Lemon -Burton, who had built a private railway in Kilburn, West London. He had it brought down by rail. However he did not keep it long, for in 1947 it was purchased by Sir John Samuel. Along with a very dedicated group of friends, Sir John had started building a 7¼in gauge railway in the grounds of his house at Burwood park, near Walton-on-Thames. Their only problem was that they had not been able to get hold of a decent steam locomotive. Louis Shaw's 4-6-2 soon changed that, and Sir John thought so much of it that he christened it EUREKA.

Later a nameplate was cast with the name, and the LMS 6200 livery gave way to No 1947 of the Greywood Central Railway, as it had become. It should be added that although Sir John believed that only two 4-6-2s had been built like EUREKA, in fact at that time he never learned of the true origin of the locomotive. Even Harry Ferguson in Carnoustie never knew that his No 6200 had been built by Louis Shaw - H-K had never told him of the locomotive's origins. I have only now been able to piece the story together.

On 26 December 1951 Sir John drove EUREKA with a two coach train around the continuous route at the G C R to see what distance he could achieve non-stop. Extra water cans were passed to him en route, and additional coal supplies were stowed in the train. It even proved possible to oil the locomotive without stopping, provided that the route was taken slowly. The result was a remarkable run of 60 circuits, each one recorded by the Official Timekeeper Mr A B Macleod. That totalled no less than 13.57 miles of continuous running.

The later history of the Greywood Central Railway is a subject all to itself. Sufficient to say here that it went on from strength to strength. By 1961 the complex layout was completely signalled and could be operated authentically using a prepared timetable; there were some 12 locos on the roster that year. Sadly, Sir John Samuel died in 1962, but thanks to the publisher Ian Allan a new site was found at Hardwick Lane, Lyne, near Chertsey. With a greater area available it was possible to build a still more ambitious layout, which evolved from 1968 as the Great Cockrow Railway. Now owned by Ian Allan, EUREKA has most recently been overhauled by Ray Hammond of Reading, and the loco is still giving excellent service today.

1947 EUREKA, at Broad Oak depot on the Greywood Central Railway, circa 1955.

1935 and PACIFIC LMS 6200 has been photographed on the turntable. There are many detail differences between this and the earlier PACIFIC, but unlike 4471 and 4472 this loco clearly has a rivetted boiler.

1937, with LMS 6200 in action at the Beach Model Railway, Carnoustie. This locomotive is now better known as 1947 EUREKA on the Great Cockrow Railway.

A fine portrait of LNER 4471, taken in 1933. Some smaller models are on display in the showcase behind.

By 1937 at Mablethorpe, Louis Shaw's last pacific, 4472, was in use. this is the only photograph showing the latter years of the High St. line in operation.

Now we go back to Louis' next locomotive. By the time of a 1929 visit, a green 4-6-2 was working on the Mablethorpe Model Railway, carrying fully lined livery as LNER No 4471, and rods painted in red. This locomotive was the first to be built by Louis in H-K's Mablethorpe workshop. It was very similar in design to PACIFIC. This locomotive is the one most often pictured at work in the various photographs; it stayed on the railway right to the end. What happened to it then is not absolutely clear, except for what I have been told by a Mablethorpe man who was later (post war) a partner with H-K. He recalled that one of the 4-6-2s which worked on the Mablethorpe Model Railway H-K kept on display inside his home at Spire House, right through to the 1950s. He finally parted with this loco. circa 1957, to a customer in the U.S.A. or Canada. It was crated up along with some rolling stock and track and everything driven down to Southampton Docks for onward shipment.

In the early months of 1936, Louis completed his fourth and final 4-6-2. This stayed close to the original design, but carried LNER livery as No. 4472 throughout its life on the Mablethorpe Model Railway. Although this loco looked similar to No 4471 above, I am convinced that in fact they were two different locos. Evidence for this is provided by the fact that in the 'Railway Magazine' of June 1937 Mr Charles H Lea clearly reported seeing two freelance 4-6-2s at Mablethorpe. I know that he wrote this from first hand because he took one of the photographs that illustrated the article. Godfrey Shaw told me that this locomotive was completed in 1936, though the first identifiable photographs I have seen of it are in 1937.

After the Mablethorpe Model Railway closed, LNER No 4472 spent the war years lying in H-K's washroom at Spire House in a dismantled state. At the 1947 sale she was sold carrying the name FLYING SCOTSMAN along with six coaches to Mr G H Waite, of Broughton Lodge Farm, at Upper Broughton on the

Leicester to Newark Fosse road. Mr Waite was indeed a colourful character. Despite being in his eightieth year, he set about the 4-6-2 with enthusiasm in between outings in 'The Old Grey Mare', a steam car which he had built back in 1906! He also managed to put down about 200 yards of track in a field next to the main A46 road. The track was adjacent to the Moneyash Cafe, and when the loco had been restored, Mr Waite opened his line as an attraction, the Broughton Lodge Miniature Railway. He affixed a nameplate to the loco, by coincidence he chose to call her LORNA DOONE.

LORNA DOONE ran at Broughton Lodge for a number of years, until the death of Mr Waite, after which it slumbered in the shed there until a chance event in May 1956. A Midlands industrialist named Michael Lloyd happened to be passing, and had paused in the cafe for refreshment, as he later recounted in the M E. Seeing a large sign outside with the words 'Model Steam Railway'. he decided to investigate, and was thrilled to find the grass grown 7¼in gauge track. "Further research discovered a neglected but serviceable-looking Pacific locomotive, complete with eight wheel tender. ..in another shed were found some passenger truck bodies, and after ten minutes puzzling, it appeared that these could be made up into two articulated sets of three cars each." The best was yet to come. Mr Waite's widow appeared on the scene, and mentioned that the whole outfit was for sale at £300. Michael wrote the cheque on the spot, and within a week the track was lifted and everything removed to a Wednesbury works for storage.

At the Hilton Valley Railway on 15 June 1969, LORNA DOONE prepares to leave Hilton.

LORNA DOONE hauls its train out of Lawn Pastures towards Hilton in June 1969.

Michael owned a house with considerable grounds at Hilton, between Bridgnorth and Wolverhampton. Within a month of purchasing LORNA DOONE work had begun, and the 'Hilton Valley Railway' opened on Easter Sunday 1957. After the first season, the loco was sent to Bagnalls of Stafford where she received a new copper firebox and a thorough mechanical overhaul. After that LORNA DOONE proved a most reliable hard worker. She ran anything up to 20 miles in traffic each Sunday, often blowing at the safety valve in spite of pulling an average load of more than one ton. She was rebuilt again by A J Glaze in 1969.

Although extensive flooding in 1960 washed away part of the track, this was soon restored, and something new was added every season on the HVR. Run by volunteers with all takings going to charity, the railway became a great inspiration to 7¼in gauge followers in the Midlands. By the mid 1960s there was a ten minute run from Hilton through Lawn pastures, Bradeney Bridge, to Stratford Brook and back. Of course there were extra locomotives as well, eight in all by the early 1970s.

When he bought LORNA DOONE, Michael Lloyd had not been able to find out who built her, but in 1970 Louis Shaw made a special outing to see her. At eighty nine years of age it must have been a very special occasion for him. He drove the locomotive which he had not seen since he finished building it back in 1936; at least he found it in good hands.

The Hilton Valley Railway closed in September 1979, and LORNA DOONE moved with the rest of the stock to Weston Park, where a new line was built. However the 4-6-2 was not well suited to the Himalayan gradients there, and in 1983 she was sold for use on the Great Cockrow Railway where she worked side by side with EUREKA for a while. In 1986 she was sold again, this time to Colin Cartwright of the Walsall Steam Railway. Colin especially wanted to own this engine as it was the first 7¼in gauge loco he ever drove, at Hilton. He had the loco

Louis Shaw with his final, masterpiece, locomotive R D LAWRIE.

R D LAWRIE with steam up, on a short line in the garden at Longfield Lane.

Strangers at Mablethorpe. Both of these locomotives were offered for sale by Harding-Kiff, photographed on the shed siding at High St. This FLYING SCOTSMAN appears to be a fine machine.

This locomotive is clearly a Bassett-Lowke IMMINGHAM.

engine as it was the first 7¼in gauge loco he ever drove, at Hilton. He had the loco on static display at the Conwy Valley Railway Museum for a while. In August 1988 LORNA DOONE was sold to Christopher Fincken, who is building an extensive new 7¼in gauge railway at his home in Herefordshire.

R D Lawrie

R D LAWRIE was Louis' last 7¼in loco, as described in chapter six. Although they have received a number of offers for it, R D LAWRIE has remained in the ownership of the Shaw family. In fact it has passed down three generations - first Louis, then Godfrey and now Brian Shaw, who is currently overhauling it.

2-4-0 Chassis

Apart from all the smaller models that he built, at the time he left Mablethorpe Louis had begun work in his Ilkeston workshop on a 7¼in 2-4-0 for H-K. The chassis, cylinders and tender were all complete and had to be handed over to H-K at Mablethorpe; these were all advertised for sale by him in the M E of 10 December 1936.

Bits and Pieces

It seems appropriate to write a few words about the other locomotives which H-K handled at Mablethorpe before the War, although I could never hope to cover them all. I do have two photographs (reproduced opposite) which H-K sent to Matthew Kerr, evidently advertising the locos as for sale. One shows what seems to be a very fine model of 4472 FLYING SCOTSMAN, which I know nothing else about. The second shows a Bassett-Lowke IMMINGHAM. This was presumably the B-L 4-6-0 which H-K advertised for sale in the M E on a number of dates in the latter half of 1935. However, the photo. itself must have been taken some time earlier, for in the background the old station building can be seen, which was replaced in 1933.

For the record I should now name all the locos which appeared on the Mablethorpe Model Railway in 1935. They were both of Louis Shaw's 4-4-2s, 4-6-2s LNER No 4471, LMS No 6200, and the B-L IMMINGHAM.

H-K managed to buy and sell at least three GEORGE THE FIFTH locos by 1947. 7¼in locomotives included in his sale that year included No 6233 DUCHESS OF SUTHERLAND built by Powell, and Midland Railway 0-4-4 No 1830, the model built by Twining in 1915. Were I ever to trace H-Ks own records of everything he bought and sold, there would be no telling what I might find!

Tailpiece

Though he himself is now gone, Louis Shaw has left behind him some fine 7¼in gauge locos, which will probably see out the present generations. I have found great pleasure in bringing together his story, and hopefully I have shared some of that pleasure with you the readers. Everybody I have met who knew Louis has told me how he himself was a polite and modest man, a fine craftsman. That is probably as good a tribute as any to finish on. Louis' death in 1971 was a great loss to the miniature railway world.

Chapter Eight Acknowledgements

From the outset I was aided by the Shaw family, especially Brian and Godfrey, without whom this booklet would not have been possible. From modest beginnings, the 'Louis Shaw Project' grew to become a real team effort. Peter Chambers was responsible for much valuable local research in Mablethorpe. It was Simon Townsend's idea that the whole story should be published from beginning to end in one booklet. Simon gave an enormous amount of time in helping to gather the material, and then to shape the manuscript for publication in this form. George Barlow and John Hall-Craggs gave me particular encouragement. Once the first draft was written, Harold Bowtell and Rodney Weaver made some extremely useful contributions to it.

The artwork for the cover was prepared by Brian Clarke, who also drew the maps. Most of the Shaw family collection photographs were copied by Keith Stratton, often from small and/or faded originals. From the 7¼in Gauge Society, Brian Rogers, Mike Taylor and Jack Meatcher gave particular help at different stages. Ron Green prepared the booklet for the printers.

I would also like to thank the following for all their assistance: Angus Tourist Board, G Anthony, Councillor Jack Bailey, Battison family, T Baxter, Chris Bishop, Councillor David Bookbinder, Wilf Boothby, Ron Bray, Derek Brough, Colin Cartwright, Lewis Coles, Percy Cooke, G Corder, John Davis, Derby Evening Telegraph, Dundee Weekly News, Mrs Evans, Chris Fincken, J Gardner, Mr Garner, Ivor Gotheridge, Grimsby Evening Telegraph, Mr and Mrs Guild, Eric Hackett, C P Hall, G Help, Dave Holroyde, Hornsea Gazette, Ilkeston Advertiser, Ilkeston Library, Ilkeston Local Historical Society, Arnold and Mrs Inglis, Mr Jobling, Matthew Kerr, Kodak News, Mr Lemon-Burton, Lincolnshire Newspapers Limited, Lincolnshire Library Services, Mablethorpe Library, Cyril Machin, D Manners, Mills family, Bob Moore, Geof Nicholson, Nottingham Evening Post, Nottingham SMEE, Nottinghamshire Library Services, John Plumpton, Stan Robinson, Mr Rogers, BBC Radio Derby, Radio Trent, Mrs Sayers, Scots Magazine, Stanley Shooter, Mr Sheffield, A Stones, Stones family, B Stuart, Keith Taylorson, Mr and Mrs Thomas, Thornton Newspapers, Mr Usher, A Wheatley, John Williams.

Last but not least I would like to thank my wife, for putting up with my researchings throughout the last three and a half years. Despite all of the above combined effort, there remain a number of gaps in the story. Perhaps some of the "lost" Louis Shaw engines will now re-emerge. I remain keen to hear from anyone who can assist further; at 8, Dale Road, Stanley, Ilkeston, DE7 6EY.

Photograph Credits

George Barlow Collection 20 Lower (L), 23; Ken Blackham 16; Derek Brough and his Collection 20 Upper (U), 40; Ken Hartley NGRS Collection (Courtesy of Ron Redman) 28U, 41; Dave Holroyde 43L, 44U; Matthew Kerr Collection 35U&L, 39U, 42U, 46U&L; Mrs Mills 39L; Mrs Sayers Collection 36U; Mr Stones 21U&L; Simon Townsend & Collection Inside Front Cover, 36L; Miss Trott Collection 5U; David Turner Inside Back Cover. All of the remainder are from the Shaw Family Collection.